Life in a
TIDE
POOL

Clare Oliver

**RAINTREE
STECK-VAUGHN
PUBLISHERS**

A Harcourt Company

Austin   New York
www.raintreesteckvaughn.com

Copyright © 2002 Steck-Vaughn Company

All rights reserved. No part of this book may be reproduced or utilized in any form or by any means, electronic or mechanical, including photocopying, recording, or by any information storage and retrieval system, without permission in writing from the Publisher. Inquiries should be addressed to: Copyright Permissions, Steck-Vaughn Company, P.O. Box 26015, Austin, TX 78755.

Published by Raintree Steck-Vaughn Publishers, an imprint of Steck-Vaughn Company.

Project Editors: Sean Dolan and Kath Walker
Production Manager: Richard Johnson
Illustrated by David Webb, Mike Atkinson, and Colin Newman
Designed by Ian Winton

Planned and produced by Discovery Books

Library of Congress Cataloging-in-Publication Data

Oliver, Clare.
Life in a tide pool/Clare Oliver.
p.cm. -- (Microhabitats)
Includes bibliographical references (p.).
ISBN 0-7398-4332-X
1. Tide pool ecology--Juvenile literature. [1. Tide pools. 2. Tide pool ecology. 3. Ecology.] I. Title.

QH541.5.S35 2001
577.69'9--dc21

2001019554

Printed and bound in the United States
1 2 3 4 5 6 7 8 9 0 LB 07 06 05 04 03 02

Acknowledgments
The publishers would like to thank the following for permission to reproduce their pictures:
Front cover: Richard Herrmann/Oxford Scientific Films; p.7: Greg Balfour Evans/Greg Evans International; p.8: David Woodfall/Natural History Photographic Agency; p.10: Charles & Sandra Hood/Bruce Coleman Collection; p.13: N.R.Coulton/Natural History Photographic Agency; p.14: Rodger Jackman/Oxford Scientific Films; p.15: Fritz Polking/Frank Lane Picture Agency; p.16: D.P.Wilson/Frank Lane Picture Agency; p.17: Kim Taylor/Bruce Coleman Collection; p.18: Richard Herrmann/Oxford Scientific Films; page 20: William S.Paton/Bruce Coleman Collection; p.22: Charles & Sandra Hood/Bruce Coleman Collection; p. 23: PhotoDisc inc.; p.24: Charles & Sandra Hood/Bruce Coleman Collection; p.25: Roy Waller/Natural History Photographic Agency; p.26: Jeff Foott/Bruce Coleman Collection; p.27: Tui de Roy/Oxford Scientific Films; p.28: Paul Kay/Oxford Scientific Films; p.29: E & D Hosking/Frank Lane Picture Agency.

Contents

At the Shoreline — 4
Tide Pool Plants — 10
There to Stay — 12
Just Visiting — 16
Tide Pools Under Threat — 28
Glossary — 30
Further Reading and Websites — 31
Index — 32

At the Shoreline

The Tide Pool

Tide pools, also called tidal pools, are found at the seashore, which is where oceans and seas meet the land. No two tide pools are alike. Some are found in hollows in the rocks and contain water almost all the time. Others are created by the action of waves on a sandy beach, but the water drains away quickly, and such tide pools exist for only a brief time.

A Harsh Habitat

Twice a day, when the **tide** comes in, the pool is flooded by salty seawater. In summer, the hot sun may dry up the pool completely. In rainy weather, the pool fills with rain, and the water becomes less salty. Since conditions in a tide pool are constantly changing, the animals and plants that live there must be tough and adaptable.

Guess What?

- Rocks are a symbol of permanence and strength.
- In China, crabs symbolize dishonesty because they move sideways rather than straight ahead.
- The crab is the symbol for Cancer, one of the signs of the Zodiac.

sea star · crab · goby · shrimp · dog whelk · limpet

High and Low Tides

At most seashores, the tide rises and falls twice a day. At high tide, pools high up on the beach are filled with foaming seawater (below).

At low tide, the sea runs out to reveal low-level pools that are normally underwater (below).

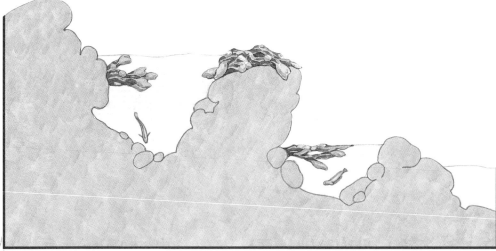

The Moon's Pull

Tides are the result of the force known as **gravity**. In this case, gravity exerted by the Moon influences the movement of water on Earth. The gravity is strongest at the parts of the Earth that are directly facing or directly opposite the Moon. Because the Moon and Earth are not still but moving through space, the area of strongest gravity keeps changing, causing the tides.

Exploring a tide pool is fun, but try not to disturb the animals.

Guess What?

- At low tide, the sea may go out as far as 2 miles (3 km).
- The fastest-moving tides happen in the Bay of Fundy, Canada. At high tide, the water can reach 69 feet (21 m) above sea level.
- A neap tide is the name for the tide with the smallest variation between high and low tide.

A Day at a Tide Pool

The animals and plants that live in a tide pool have to adapt to the changes that result from the changing tides. At low tide, they are more exposed to the Sun or to the cold of the night—and to **predators**! To protect themselves, soft-bodied shellfish close their shells, or burrow down into the sandy bottom of the pool.

Constantly changing conditions brought about by the tides make the tide pool a unique habitat.

High Tide

High tide brings new difficulties for life in the pool. Shellfish anchor themselves firmly to the rock to withstand the battering of the waves and avoid being swept out to sea. High tide is feeding time, as the tide brings fresh **plankton** from the sea. Barnacles extend feathery legs and anemones wave their tentacles to pick up their meal.

Guess What?

★ Sea stars die out of water, so they cling to the base of a rock with their hundreds of tiny feet.

★ Barnacles make superstrong "cement" to fix themselves to the rock. The "cement" sets underwater and will even stick plastics together!

★ A mussel uses its beard, or byssus, to grip the rock.

Barnacles open their plates at high tide. They poke out feathery legs to pick up plankton swept in by the sea.

The barnacles close up again at low tide.

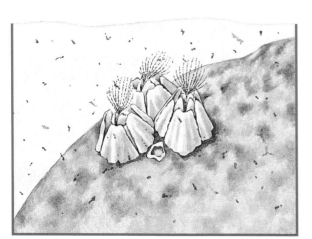

Tide Pool Plants

Seaweeds

Tide pools at the foot of the beach are nearly always covered by water. Slimy brown seaweed called kelp grows here, where it will not dry out. Higher up the beach, tide pools contain smaller seaweeds, including wracks. Instead of having roots, seaweed clings to the rock with a tough stem called a holdfast.

Seaweed is not really a plant but a type of organism called **algae**. It can absorb nutrients (food) from the water through every part of its surface.

A Place to Hide

Seaweeds provide protection from the Sun and the winds for small creatures such as fish, crabs, and mollusks. Hiding among the fronds is a good way to avoid being gobbled up by a gull, too! Seaweeds help animals in other ways—by producing oxygen for them to breathe or as the source of a tasty meal.

Bladder wrack

Sea lettuce

Irish moss

Coral weed

See for Yourself

This kelp has scars where it's been nibbled by tide pool creatures.

The pouches on this piece of bladder wrack are filled with air. They act as life buoys, so the seaweed floats!

There to Stay

Clinging On

Many tide pool creatures have shells to protect them from attack by predators.

Common periwinkle

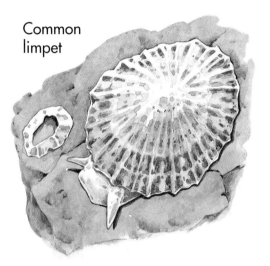
Common limpet

Most shellfish belong to the group of animals called **mollusks**. Inside their hard shells they have soft, squishy bodies.

The ones that have shells with twin halves —such as mussels, clams, scallops, and rock oysters —are called bivalves.

Variegated scallop

Mussel

Mollusks with a single-piece shell are called gastropods. They include snails, whelks, limpets, and periwinkles. Dog whelks are predators, but most mollusks graze on seaweed and other algae. Others sift through microscopic food in the water.

Periwinkles have hard, rounded shells to protect them from predators and the heat of the Sun.

See for Yourself

The color of a dog whelk's shell says a lot about its diet. Yellow dog whelks eat pale-colored barnacles, while purple ones eat mussels.

This "path" on the rock shows where limpets have been grazing on seaweeds and other algae.

Scary Stingers

Sea anemones are beautiful tide pool creatures that come in all the colors of the rainbow. They are also scary predators, feeding on tiny fish, shrimps, and marine worms.

The elegant, waving tentacles of the sea anenome are armed with a painful sting to ward off predators or to stun **prey**.

Anemones belong to the same family as the wobbly jellyfish that live out at sea or are sometimes washed ashore. Like jellyfish, they have very simple bodies.

Dinner Time!

When an anemone catches a small shrimp or worm, it pulls it in toward its mouth, at the center of the ring of stingers. The meal is digested in the anemone's small stomach.

In this picture you can see the anemone's mouth surrounded by tentacles.

Guess What?

★ Sea anemones are named after flowers. Long ago, people mistook their waving tentacles for petals.

★ A sea anemone can divide itself in two to make two new complete creatures, but some produce tiny copies of themselves that swim out of their mouth!

★ When it pulls in its tentacles at low tide, the strawberry anemone looks just like a juicy strawberry.

Just Visiting

Tiny Creatures

Thousands of tiny life forms swish in and out of the tide pool with the rising and falling of the tide. This soup of floating plants and animals is called plankton.

The plants are known as phytoplankton and include various algae, such as diatoms. Zooplankton is the name for the microscopic animals, eggs, and larvae (young) found in the seawater.

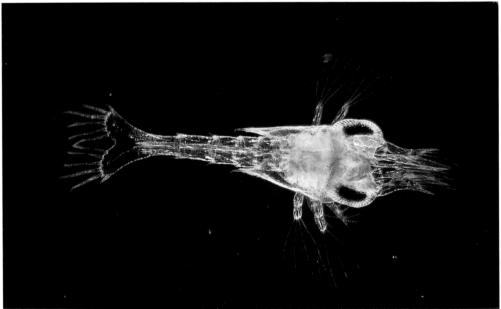

Lobster larvae are only about one-fifth of an inch (5 mm) long when they hatch!

Watery Nursery

Some of the eggs found in tide pools come from creatures such as shrimps, crabs, lobsters, or dogfish that breed in shallow waters. When the larvae hatch, they live in the low-level tide pool, until they are ready for life out at sea. Other creatures spend their whole life in the pool.

This shrimp is carrying its eggs in an egg sac.

See for Yourself

This "mermaid's purse" was a dogfish egg case. The baby fish developed in here for about 10 months before it hatched.

This pink string is the spawn of a sea hare, an ocean creature that looks a lot like a sea slug.

These are dog whelk egg cases. When they hatch, dog whelks look exactly like miniature versions of their parents.

Rock Stars

Lots of different types of sea stars, or starfish, hunt in the tide pool. They range from cushion stars, which have plump bodies and rays (arms), to brittle stars, which are spiny and spindly.

At low tide, sea stars hide out under the rocks, but at high tide they come out to find mussels and other mollusks to eat.

Making a Meal of It

Sea stars have a strange way of eating mussels! First, they wrench open a gap in the mussel's shell with their strong rays. Then they turn their own stomach inside-out, pushing it out of their mouth and into their victim's shell. When it's finished digesting the meal, the sea star swallows its stomach back down again.

Guess What?

- The sea star's eyes are on the end of its rays. Their eyesight is not like ours though, because they can only sense light and dark.
- Not all sea stars have five rays. Some have more than 50!
- When a ray breaks off, the sea star can grow a new one.
- Baby sea stars are called starlets.

Common sun star

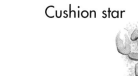

Cushion star

Brittle star

Prickly Customers

The sea urchin's sharp spines provide good protection against its enemies—as you would quickly find out if you accidentally stepped on one! Its rounded body is covered with movable spines that, together with hundreds of tube-like feet tipped with suckers, help the animal to move about and even to climb rocks.

The sea urchin's mouth is on the bottom of its body. It has five strong teeth which it uses to scrape algae and tiny mollusks off seaweed and rocks.

Sea urchins graze on algae, tiny mollusks, and seaweed, but they will also eat the dead flesh of other animals when available.

Strange Symmetry

Underneath the forest of spines and below the skin is a hard outer skeleton called a test. When a sea urchin dies, its spines break off and you can see the test in all its beauty —you may have found one washed up on the seashore or seen one on sale at the seaside.

A sea urchin's test showing a section covered with spines and legs.

- Tube feet
- Small holes where feet project.
- Spines
- Raised bumps to which spines attach.

Guess What?

The beautiful shell-like tests of sea urchins are often sold as ornaments for the home.

The teeth and jaws of a sea urchin together are known as Aristotle's lantern, because the shape is like that of an old-fashioned lamp.

Humans are also predators of the sea urchin. Some consider the **roe** (eggs) to be a delicacy.

Some sea urchins can regrow their spines when they break off.

There are hundreds of bumps on the test to which the spines were attached and rows of holes through which tube-like legs once extended.

Crabs and Lobsters

As they scuttle between the rocks, crabs are the tide pool creatures that are perhaps the easiest to spot. They are members of a class of animals known as **crustaceans**.

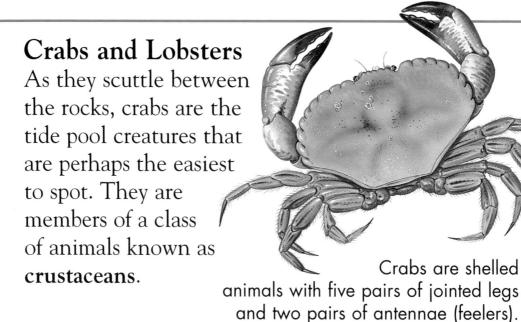

Crabs are shelled animals with five pairs of jointed legs and two pairs of antennae (feelers).

Lobsters (below) and crayfish are crustaceans, too. These animals act like the tide pool's garbage collectors, scavenging for decaying bodies on the bottom of the tide pool.

Moving Home

When crabs and lobsters grow too big for their shells, they molt, shedding the old shell to reveal a shiny new one underneath. Hermit crabs (below) are different. Only the front part of their body is plated—they look for an old snail shell to protect the rest of their body and move into bigger ones as they grow.

Guess What?

Pea crabs are the tiniest crabs. You might even have found one on your dinner plate, because pea crabs live in the shells of mussels and oysters.

Green crabs aren't necessarily green! Males can be such a dark green they look black, while females and the young come in all manner of patterns and colors.

Crustaceans may have up to 17 pairs of legs.

Lobsters have blue blood.

Smooth Swimmers

The tide pool habitat is too small, shallow, and variable to be a suitable home for most fish.

Tide pool fish like this blenny are usually between 2 and 8 inches (5 and 20 cm) long.

Only small types, such as blennies, gobies, and sea scorpions live there all the time, feeding on small shrimps and other creatures. Even so, larger fish are sometimes swept in on the surf and stranded in the pool between tides.

Special Features

Gobies, sea scorpions, and blennies have adapted to life in the tide pool. Their eyes are positioned high on their head so they can check above for the shadow of a predator. Their mottled skin blends in with the rocks and pebbles, but they can also swim very fast. They dart into cracks in the rock if they sense danger.

Sea scorpions can turn lighter or darker for better **camouflage**.

Guess What?

Sea scorpions lay their bright orange eggs at the bottom of deep tide pools. The male guards the eggs until they hatch.

Blennies do not have scales, so their bodies are much smoother than other fish.

Sand gobies burrow into the sand to hide from predators.

Feathers and Fur

All sorts of predators stop by at the tide pool for a meal, especially hungry birds, such as gulls, eider ducks, and shore birds. They pick through the seaweed and stones at low tide, in search of crabs, mussels, and other food to eat. If a shell is too tough for its strong beak to crack, a gull might drop it from a great height onto the rocks below, so that it smashes.

The tide pool is a rich source of food for the strong-beaked gull.

Sunbathing

Many animals like to sit on the rocks around the tide pool, basking in the Sun. On the Pacific Coast of the United States, these include seals and sea lions (above). These animals spend most of their lives out at sea but come ashore in huge numbers to give birth to their young.

Guess What?

Eider duck feathers are collected to make beautifully soft eiderdown quilts.

Can you guess how the bird known as the turnstone gets its name? These shore birds turn over stones in search of small crabs, using their flat beaks.

Although they are born on land, seal pups can swim almost as soon as they are born.

Tide Pools Under Threat

Changing Tide Pools

The force of the waves wearing on the rocks and sand is constantly changing a tide pool, but this change takes place very slowly. Human activity changes the coastlines, too. Global warming—a rise in Earth's temperature—could lead to higher sea levels in years to come, leaving today's tide pools under water. Pollution is another result of human activity that harms tide pool organisms.

Garbage left behind or brought by the sea can destroy the delicate balance of the tide pool microhabitat.

Polluted Pools

Estuaries are the place where rivers meet the sea. Some become dumping places for dangerous chemicals from factories and farmland or sewage from our homes. These poisons are absorbed by tiny plankton, then enter the bodies of larger animals that feed on plankton, such as mussels. In this way pollutants can travel right up the **food chain** into bigger and bigger animals.

Guess What?

The worst shoreline oil spill happened in the Gulf of Alaska in 1989, when the oil tanker Exxon Valdez ran aground and spilled almost 11 million gallons (42 million liters) of oil along 1,300 miles of Alaska's southern coast.

Herring gull populations have increased in recent years. This is because they feed on sewage emptied into the sea from pipes on land and ships at sea.

If global warming continues at its present rate, sea levels could rise as much as 2 to 4 feet (60 to 120 cm) in the 21st century.

Oil spills are another problem—when huge tankers carrying crude oil leak or run aground, all life along the shore suffers and some species disappear forever.

Glossary

Algae: Living things, such as seaweeds, that live mainly in water. Like plants, algae make energy from sunshine, but they are not plants because they do not have leaves, stems, or roots.

Camouflage: Coloring, or a means of disguise, that makes an animal blend in with its surroundings so that it is more difficult for predators to see.

Crustaceans: Armor-covered animals such as crabs. They have jointed legs, like insects, but only two body parts, not three. Most crustaceans live in water.

Food chain: A series of plants and animals in a microhabitat that are linked because each becomes food for the next one in the series. Large predators seem to be at the top (or end) of the food chain, but when they die they rot down and feed new plants or algae that are nearer the bottom of the food chain.

Gravity: The force that pulls an object toward a heavier object. Gravity pulls the Moon toward the Earth and the Earth toward the Sun.

Mollusks: Boneless animals with soft, squishy bodies that need to be kept damp and are sometimes protected by a shell. Snails and squid are both types of mollusks.

Plankton: A soupy mix of plants and animals that are so small they can only be seen through a microscope.

Predators: Animals that hunt other animals for food.

Prey: Animals that are hunted by other animals for food.

Roe: The mass of eggs inside the body of some aquatic creatures.

Tides: The regular rise and fall of the sea.

Further Reading

Berger, Melvin and Gilda Berger. *What Makes an Ocean Wave?: Questions and Answers About Oceans and Ocean Life.* New York: Scholastic Trade, 2001.

Bredeson, Carmen. *Tide Pools.* Danbury, CT: Franklin Watts, 1999.

Goodman, Susan E. *Ultimate Field Trip 3: Wading into Marine Biology.* New York: Atheneum, 1999.

Ocean Life (Time Life Student Library). New York: Time Life, 1999.

Steele, Philip. *A Tide Pool.* New York: Crabtree Publishers, 1999.

Websites

http://www.umassed.edu/public/kamaral/thesis/tidepools.html
Learn how to cook seaweed pudding (with the help of an adult) or click on "barnacle" and watch a barnacle eat!

http://www.pbs.org/wnet/nature/edgeofsea/index.html
Learn about life in a tide pool, enjoy the colorful pictures, and take a virtual tide pool trip during high tide or low tide.

Index

algae 10, 13, 16, 20
Aristotle's lantern 21

barnacles 9, 13
birds 26
bivalves 12
bladder wrack 11
blennies 24, 25
brittle stars 18, 19

camouflage 25
chemicals 29
clams 12
common sun star 19
coral weed 11
crabs 11, 17, 22–23, 26, 27
crayfish 22
crustaceans 22, 23
cushion stars 18, 19

diatoms 16
dog whelks 13, 17
dogfish 17

eggs and egg cases 16, 17, 21, 25
eider ducks 26, 27
estuaries 29

fish 11, 14, 24–25
food chain 29

garbage 28
gastropods 13
global warming 28, 29
gobies 24, 25
gravity 7
green crabs 23
gulls 26

hermit crabs 23
herring gulls 29
holdfast 10

Irish moss 11

jellyfish 14

kelp 10, 11

larvae 16, 17
limpets 12, 13
lobsters 16, 17, 22–23

"mermaid's purse" 17
mollusks 8, 9, 11, 12, 13, 20
molting 23
Moon 7
mussels 9, 12, 19, 23, 26, 29

oil spills 29
oxygen 11
oysters 23

pea crab 23
periwinkle 12, 13
phytoplankton 16
plankton 9, 16, 29
pollution 28, 29

rock oysters 12
roe 21

sand gobies 25
scallops 12
sea anemones 14–15
sea hare 17
sea lettuce 11
sea lions 27
sea scorpions 24, 25
sea stars 9, 18–19
sea urchins 20–21
seals 27
seaweed 10–11, 13, 20
sewage 29
shellfish - see mollusks

shrimps 14, 15, 17
snails 13
starfish - see sea stars 18
strawberry anemone 15
Sun 8, 11, 27

test 21
threats to tide pools 28–29
tides 5, 6, 7, 8, 9, 16, 24
turnstones 26

whelks 13
worms 14, 15
wracks 10

zooplankton 16

```
J 577.69 OLI
Oliver, Clare.

Life in a tide pool.
Raintree Steck-Vaughn,
c2002.
```

JAN 2003

Glen Rock Public Library
315 Rock Road
Glen Rock, N.J. 07452

201-670-3970